WRITING
in earth science

Robert L. Bates

American Geological Institute
Alexandria, Virginia

Published by the American Geological Institute

Printed and bound in the United States of America

ISBN 0-913312-92-4

TABLE OF CONTENTS

AUTHOR'S NOTE

Sooner or later, every young earth scientist will be faced with the necessity of writing a report. This may be a senior thesis or a doctoral dissertation, an inter-office memo or a recommendation for company action. The prospect of having to commit words to paper, so as to convey a message in a logically organized form, is likely to produce a variety of reactions, few of them pleasant. Because I have been both editee and editor of geological reports, and sympathize with those who must make the written language work for them, I have prepared this book. It is intended for young earth scientists who may be sure of their science but are not so sure of their ability to communicate it.

The object of your writing must be to communicate something from your mind to the reader's mind, clearly and accurately, with as little static as possible. It follows that the whole process must be reader-directed. "What does this say to the reader?": that is the question. The English language, an enormously adaptable means of transmitting a "cerebral itch" from one mind to another, is at your disposal.

In this book I present rules and recommendations on such matters as sentence structure, acceptable forms of expression, and terms that should be used with precision if used at all. Many items in these categories seem to trouble beginning writers in earth science. Perhaps the bad examples cited will encourage you to try to do better than others have done. In any case, mastery of the precepts set forth will help you avoid much editorial grief. It will also give you additional confidence and polish in your use of the language.

At a 1985 meeting of biology editors, a speaker discussed the electronic "manuscript," which the author submits to the publisher as a computer disk. Drawbacks include the large number of incompatible systems and the difficulty of digitizing and publishing photographs. No doubt these problems will be overcome. But the speaker acknowledged a third drawback: the great editorial responsibility this system places on the author. Note that no provision appears to be made for editing. What you type on your word processor is what will appear in the journal. In this brave new world, a greater premium than ever must be placed on being able to write well.

Robert L. Bates
February 1988

CHAPTER 1

THINKING

Three questions need to be answered at the very beginning of your writing project. These relate first to your audience and your message, and then to your title.

Who is your audience?

For students preparing their senior theses, the answer may be fairly obvious: they are writing to satisfy the professor who is their adviser. Such theses are on a modest scale and are unlikely to reach a wide audience. But master's theses and doctoral dissertations are on more lofty planes and will no doubt be considered by a committee. Who will be on this committee, and what will they expect? Your adviser will be acquainted with your material, but how about the others?

A thesis or dissertation may warrant publication. In fact, some professors require that any dissertation prepared under their direction be in the format required by a specific journal. Or, if you present your report at a meeting, it may appear in the published proceedings. There are many other possibilities. In any case, you need to have a good idea of who will read what you write.

Audience-consciousness is fully as important if you are a company geologist. You had better know ahead of time whether your report will be read only by your supervisor, or whether it will be passed up the chain of command. Who will approve the report, or act on it? You might even request a pre-writing conference, in order to answer these and related questions.

The manufacturer of a household product gives much careful thought to the wishes and needs of the ultimate consumer. You should do no less.

What is your message?

Of course you know in a general way why you are writing the report. But "in a general way" ought to be refined. What, specifically, do you have to contribute?

Most technical and scientific papers provide readers with information they did not have before: solution of a problem, let us say, or new information that may lead to such a solution in the future, or a new method of attacking an old problem. Whatever the subject, you must have the object of the whole exercise clearly in mind. Few things are more frustrating to amateur writers than finding they have spent hours or days working up material that is either marginal or even extraneous to the central message. As in tennis, keeping your eye on the ball is essential.

Whereas the scientific paper gives something to readers, the company report may request something from them, or recommend action by them. Here of course you must keep in mind just what is being requested or recommended.

Reports seldom go directly from the writer to the reader; most are edited somewhere along the line. All sorts of hassles between author and editor can be avoided if you know from the start who your audience is and what you have to tell them.

A LABEL FOR THE PRODUCT

Once you're satisfied as to audience and subject, the logical next step is to decide on a title. The first function of the title is to help you, the writer, as a guide to be kept in mind all through the writing. It defines the subject and sets the limits. Regular reference to the title prevents you from writing off into the sunset on some interesting but irrelevant topic. But the main purpose of the title is to help others. Your title will go into bibliographies, and its key words into data banks. Potential readers will get their first impression of your work by what you have chosen to call it.

To work out a title that is clear and reasonably brief seems like a simple assignment, but it is not always easy. Brevity is a goal, but it may perhaps be overdone. A paper in the geophysical literature is titled Q. Another is f_{max}. These are clear if you happen to know that Q is a seismic attenuation factor and that f_{max} is the high-frequency band limitation of the radiated field of earthquakes.

Most of us, however, have to use words rather than letters or symbols, and many titles are far more wordy than necessary. Consider, for example, this title of a paper given at a symposium: *Paleogeographic Implications Pointing to Fondo Origin of the Chattanooga-New Albany-Ohio Bituminous Shales*. If we analyze this a bit, we find that *origin* is the central or controlling word. Now if the writer discussed the origin of the shales, he must have talked about paleogeography; the two are inseparable. Therefore, *Paleogeographic Implications of* may be taken for granted and dispensed with. Second, why not use *of* rather than *Pointing to*? Third, *Bituminous* is hardly necessary, as persons attending the symposium would know the type of shales being discussed; indeed, the named shales are widely known among North American stratigraphers as being black or bituminous. Recasting, we produce *Evidences of Fondo Origin of the Chattanooga-New Albany-Ohio Shales*: 11 words instead of 16.

But further streamlining is possible. The correlatives of the Chattanooga Shale are well known, especially to those who would be likely to hear the paper presented or read it later. Why not delete the introductory phrase *Evidences of?* (What else would he be talking about?) We then end up with a succinct 6-word title that says it all: *Fondo Origin of the Chattanooga Shale*. As it happened, the title of a parallel paper given at the same symposium was *Shallow-water Origin of the Chattanooga Shale*. Had the "fondo" title been shortened as suggested, anyone attending the symposium would have known instantly that he was in for an interesting debate. He would not have had to read 10 unnecessary words to find that out.

As for clarity, perhaps you'd be surprised at the way some reports are labeled. *An Unsolved Problem—Opaque Petrological Differences between Tertiary Dykes and Lavas* tells us that the problem is opaque differences, whatever they are. *A Method for Taking Back Reflection Oscillation Photographs of Single Crystals* seems to say that the author wanted to take back his photographs. It turns out that something called back-reflection geometry is involved. A single hyphen, as in the preceding sentence, would have saved the title from absurdity. *Boring Late Precambrian Organisms* was the object of considerable hilarity, and no wonder. Apparently it occurred to neither the author nor the editor to transpose the two modifiers: *Late Precambrian Boring Organisms* puts the adjective with the organisms where it belongs. *The Fate of a Fine-grained Dredge Spoils Deposit in a Tidal Channel* seems to be about a fine-grained dredge--or even about how such a dredge spoils deposits. In the title *Active Fault Mapping and Evaluation Program*, what is active--the fault, the mapping and evaluation, or the program? By way of contrast, it is refreshing to come across a title that is not only brief and to the point, but also a bit unconventional: *Imperial Mammoth and Mexican Half-ass from near Bindloss, Alberta*.

It is not advisable to tell everything in the title, especially because such a procedure is likely to involve heaping up nouns and adjectives. In the title *Staurolite Zone Caradoc (Middle-Late Ordovician) Age, Old World Province Brachiopods from Penobscot Bay, Maine*, the subject, brachiopods, is reached only after 10 preliminary words, and the title includes locality, provenance, geologic age (twice), and degree of metamorphism. It might have been well to leave some of this information for the abstract. *Ordovician Brachiopods of the Old World Province from Penobscot Bay, Maine* moves the subject word to second place instead of eleventh and avoids the string of modifiers.

Another example of the tell-all title is *Feasibility Study of Geochemical Sampling of Streams Reaching Coast of Arctic Islands by Helicopter from DOT Icebreaker*. Not only is this excessively long, but those streams reaching the coast by helicopter give one pause. A fairly drastic reduction would give *Geochemical Sampling of Arctic Streams by Helicopter*. Other versions are of course possible. As a bit of practice, you might try recasting this title to a more readable form: *Heavy Mineral Magnetic Fraction Stream Sediment Geochemical Exploration Program*.

If you are very confident of your results, and if your editor will allow it, you may put your title in the form of a statement: *Ochoa Is Permian*, for example. Or--again with editorial approval--you may use a question, as in *How Do Hollow Concretions Form?* Obviously the paper answers the question. Certain other titles in this form are less advisable. *Does Inorganic Carbon Assimilation Cause ^{14}C Depletion in Deep-sea Organisms?* puts the author in the position of requiring seven pages and 54 references to establish that the answer is Yes. (There is in the literature at least one abstract, which, in its entirety, consists of the word *Yes*, and another of *No.*) The whole procedure seems--shall we say--a bit questionable.

Chapter 2

ORGANIZING

Once you've got the title in place, the temptation is to start writing. When you're writing, you're actually *doing* something: punching out those words gives a sense of accomplishment. Nevertheless, put off the writing until you have decided on the topics to be covered and the order of presenting them. Making an outline isn't as much fun as typing, and besides it may make your head hurt; but it can't be avoided. Writing without an outline is like hiking across unfamiliar country without a map.

Imagine that you are talking about your project with one or two interested persons; your professor, perhaps, or your supervisor, or a couple of professional colleagues. What questions will they be likely to ask? List these questions, and beside them put your answers--in conversational style, as you might express them at the time. These answers will provide a preliminary or topic outline of your report.

Questions to be answered might include:

Why did you work on this problem?

What did you find out?

How did you tackle it?

How do you know your results are valid?

How do your results fit into the big picture?

Is further work needed?

Each of these questions begs additional ones. By writing your answers you will expand your outline to a useful, detailed format for preparing the report. Before starting to write, you might do well to submit your expanded outline to the professors on your committee, or to your management group, perhaps with a request for a pre-writing conference. Thorough exploration of the field to be covered, with your specific audience in mind, will make the writing infinitely easier.

THE ABSTRACT

A conclusion implicit in the preceding section is that readers are not interested in a "suspense format" that buries the message under Conclusions at the end of the report. Suspense may be all right in a detective story, but it is undesirable in a technical or scientific paper. Readers want the news up front. This news must be given first in the abstract at the beginning of the paper. Most journals require an abstract; so do business executives, who need to have the results of a company report in quickly readable form.

The following abstract, with irrelevant words omitted, appeared in a geophysical journal:

> A method is formulated... The displacements are computed... and a principle is applied. . . The inversion is accomplished. . . The growth of the error is evaluated... The viscoelastic uplifts are calculated, and the time-dependent uplifts are computed. Finally, functions are integrated and uplifts are compared.

This abstract is noteworthy for telling nothing about what was found out. It tells what was done, but it contains no other news. The accompanying paper does contain two news items, but these are given some 8 pages later, under Summary and Conclusions. Abstracts of this type are non-informative and should be avoided. You may state briefly what you did, but the main point is what you discovered. If you find yourself writing an abstract exclusively in the passive voice ("was measured", "is discussed"), remind yourself of the readers. They want your results.

In contrast, an informative abstract conveys the message, with enough supportive detail to make it understandable. An example follows.

> During 1982 and 1983, two ground-penetrating radar surveys were carried out in conjunction with archaeological investigations in Canada. The first survey was a detailed, high-resolution radar survey at the site of a sixteenth century Basque whaling station on the Labrador coast designed to locate the graves of the Basques. The second was a rapid, low-resolution reconnaissance survey as part of a prehistory impact assessment program at the site of the new National Museum of Man in Hull, Quebec. Both surveys were experimental and were designed to see whether ground-penetrating radar would be useful for identifying and locating anomalies of archaeologial significance. Radar was successful in detecting archaeological anomalies several metres in size at both locations, and the high-resolution survey was moderately successful in identifying Basque graves. Ongoing work involves comparing radar results

with the archaeological investigations to increase the understanding of how radar can be applied to archaeology and to improve interpretation of radar responses to artifacts.

In this abstract of about 175 words, the author states (a) what was done, (b) its purpose, (c) the results, and (d) how it contributes to future work. The reader is given a clear overall view of the subject.

Many more people will read your abstract than will read the entire paper. If you present the paper at a meeting, the abstract will appear in the program and possibly in a pre-meeting issue of the society's journal. It may also appear in trade journals and in publications of abstracting services. It is therefore prudent to devote good effort to this brief feature of your report.

The abstract is normally written after the paper is completed. But suppose you have offered to present your results in a paper at a society meeting some six months hence, and it has been accepted. Your pleasure at this news is somewhat diminished when you find that an abstract must be sent in 2 1/2 months before the meeting. Your time schedule is such that you can write the paper only in those 2 1/2. months. So you're in the unhappy position of having to furnish 150 to 250 words summarizing a paper that is not yet written. This is really not so serious a problem as it seems at first. By the time you must prepare the abstract, your research will be at or near completion, and the main results reasonably clear. Your early abstract should tell briefly what you found, and should be general enough to allow room for later maneuvering if necessary. If some last-minute research result invalidates something you've said in the abstract, state that at the time you give the talk. You can then update the abstract for the final published version of the paper.

As the early abstract is to appear in the program booklet of the meeting, it is likely that you will be asked to furnish "camera-ready copy," probably on a special form. This means that the typescript that you send in will be photographically reproduced exactly as you submit it. Therefore it had better be grammatically correct and neatly typed; misspellings and other errors will be preserved for all to see.

THE INTRODUCTION

The introduction orients the reader. It should give the nature and scope of the problem, your purpose in working on it, and the methods you chose in attacking it. It should include a statement of the results, and some indication of how these add to previous knowledge--that is, how they fit into the big picture. In a company report that recommends a course of action, this recommendation should be made briefly and clearly. Simple, straightforward language, such as you might use in an oral explanation, is preferred; but, as always, keep your readers in mind. If you are writing on a highly technical subject for specialists in the field, your language and terminology will be modified accordingly.

You may find it advisable to include a brief survey of previous work on the problem, or a history of the project. Usually it is not necessary to give an exhaustive discussion, enumerating everything that has ever been done, but you might wish to summarize the conclusions of earlier workers, with references to the literature. In a company report, this would be the place for a brief account of the project to date, and an explanation of why this particular problem arose.

METHODS AND PROCEDURES

Here, obviously, you show the way you decided to attack the problem and the techniques used. Explain why these methods and techniques were chosen, tell how well they worked, and give an estimate of their reliability. If you introduced new or modified procedures, explain why and evaluate them. Your information should be precise enough so that a qualified investigator could test or duplicate what you did. In other words, your results must be reproducible.

If your techniques are conventional, of a kind familiar to your readers, this section can be relatively brief. But complex or highly specialized techniques, especially in a new application, should be explained. Unusual or unique items of equipment may be shown in illustrations. It is important to be precise, for example in such matters as localities from which samples were obtained and the analytical methods used in working on them. If you have a large amount of supporting material, such as statistical data, modal analyses, or measured sections, you might consider placing it in an appendix.

In a company report, perhaps under the heading Supporting Data instead of the one above, give the reasons for any action recommended; that is, show specifically how you arrived at the recommendation. This section would be a crucially important part of such a report.

RESULTS

Although you have given a brief statement of results in the abstract and again in the introduction, a full report is now in order. Show how your results follow logically from the information developed in the investigation. Point out any secondary or unexpected results that may have appeared. If your work failed to show something that you expected it to show, don't hesitate to say so. This section of the report may be relatively brief, but it must be clear, as it is the most important part. Don't get involved here with a discussion of significance; this will come in the next section.

Quite likely you will need to support your text with tables, graphs, maps, photographs, or other illustrations. Preparation of these materials is discussed on a later page. A word of caution: illustrations should supplement, not duplicate, text material. Choose them with this in mind.

DISCUSSION

At this point it would be well to think again of your readers. They might reasonably be expected to ask, "All right--what does all this mean?" Does your work contribute to the solution of a problem, or solve one? How do the facts or relationships you have discovered relate to others already known? What, in short, do they contribute to the big picture?

It is easy to imagine a reader asking these questions, and it should not be too difficult to answer them. But stick to the subject. The heading Discussion is not an invitation to lengthy, vaguely thought-out conjectures or hypotheses. Your job is simple: to present your conclusions in such a way that the reader will understand them and appreciate their significance.

REFERENCES

Almost certainly you will find it necessary to refer to previously published work, the references to be listed at the end of the report. Two warnings are in order here. First, citing references is an operation replete with chances for error. Double check each reference, to be sure of the author's name and initials; date of publication; title of paper; name of the journal or other publication; and page numbers. It is wise to keep a card file, each card bearing a separate citation fully written out. Abbreviations, or changes in order of items, can be made as desired for your final list. A reference to a book or other separate publication should include the place and name of the publisher.

Second, there is no universally accepted order of giving the items in a reference. (One bibliographer, who surveyed 52 scientific journals, found that 33 different styles of reference are used, and that there are 2,632 possible ways of setting out items in reference lists!) Each journal has its own style, and the editor is likely to be unpleasant if your paper doesn't follow it. Therefore it is necessary to obtain a style guide from the journal, or, if one is not available, to obtain a current copy of the journal and follow its format. If you are writing an independent report, not for publication in a standard journal, develop your own style of citation and be consistent. Abbreviations give the most trouble. Is *Journal* written out, or is it shortened to *Jour.* or *J.*? Does *American* come out as *Amer.* or *Am.*? A minority of bibliographers maintains that giving full names of journals (no abbreviations) is more helpful to the reader and takes up little additional space; but editors are convinced that their precious abbreviations save space and therefore money. Follow whatever style is required, be consistent, and be accurate.

Most geological journals list references alphabetically by author. This is an easy system for author and reader alike. Citations in the text then refer to a paper by author and year (Snarf, 1966); page number may be included if this will help the reader (Snarf, 1966, p. 215). Another method of listing references, widely used in the literature of chemistry, is by numbering the references, either alphabetically or in the order that they appear in the paper. This makes text citations shorter: (Smith, 1966) might simply be (17); but if, as commonly happens, a new reference turns up in the course of writing, it must be inserted in the series and all higher numbers must be changed. This is a tedious procedure and greatly increases the opportunities for error.

Many papers are by more than one author. All the authors are usually named in the list of references. Take care to see that all the names are spelled correctly! (Fortunately, multiple authorship in geology has not yet reached the level found in some other sciences. A recent paper in physics cited an earlier one by 127 authors, and all 127 were listed.) Citation in the text may mention both authors if there are only two (Smith and Jones, 1968), or only the first author, followed by *et al.* (meaning *and others*) if there are three or more: (Smith et al., 1974).

ACKNOWLEDGMENTS

A general rule is to keep your acknowledgments brief and in the simple declarative mode. It is not necessary to go on effusively about profound gratitude for important contributions. A bare-bones version might go something like this: "Dr. Smiley Snarf suggested the problem and provided guidance throughout the work. Bill Jones assisted me in the field and Betty Smith prepared the line drawings. The XYZ Company gave permission for reproduction of Figures 2 and 3. The work was partially supported by grant no. 12345 from the National Science Foundation."

Chapter 3

WRITING

The day has come when your professor says, "All right, your research is well in hand and your results seem valid. Now write it up." Or, your boss says, "Well, this proposition sounds reasonable to me. I think they might like to consider this upstairs. Prepare a report for management, will you? And make it good."

Panic? Not called for. Dread? Maybe at first, but should dissipate. A feeling of being inadequate for the job? Perfectly normal. But curable.

The first line of action against graphophobia, or fear of writing, is, naturally enough, to write. However you do it--pencil and paper, typewriter, word processor--put down some words, and follow them by other words. Choose your subject: a description of your field area, a summary of previous work on your topic, or an explanation of why the project is of interest. Never mind trying to achieve a formal style--just write as though you were answering orally some of the questions that an interested friend might ask. You will probably find that writing, while by no means easy, doesn't hurt as much as you thought it would. If your schooling included some good instruction and practice, you are fortunate, and this assignment should be relatively easy. But do it anyway. The way to learn how to write is to write!

We live in a permissive age. (Webster's unabridged dictionary says ain't is "used orally in most parts of the U.S. by many cultivated speakers." How they arrived at that conclusion ain't clear.) In spite of this relaxed attitude, however, certain grammatical requirements exist, especially for serious writing. In this book you will find rules--or, where disagreement exists, recommendations--on how to handle a variety of problems that may arise in descriptive or explanatory writing. As questions come along, you may wish to consult these guides to usage. Use the index. You will also find some rather spectacular examples of how not to do it.

Let us assume that you have now written several pages, encountered problems, corrected them, and rewritten. In their second or third draft, these pages have cost you a lot of effort, and, in all modesty, you believe you've got the start of a fine report. Then comes the next line of action. Ask someone whose opinion you respect to read and criticize what you have written. If you're a student your professor will fill that role; let's hope he will do it conscientiously. Or ask a colleague who has done some edited writing--indeed anyone in whose judgment you have confidence. Subjecting your painfully crafted prose to a critical eye will not be easy, but it's essential. You will get comments like "You've used the wrong word here."--"You've already said this."--"I don't get your meaning."--and so on. Smother your resentment that this person doesn't find your writing crystal clear, be grateful for the second opinion, and profit from it. A friendly but rigorous critic is a pearl beyond price.

The language is the only natural resource that can be mined without depletion. It awaits your pleasure.

FORMAT AND STYLE

If you're writing a thesis or dissertation, the graduate school will notify you of their requirements as to typescript, number of heads and subheads that are acceptable, style of abbreviations, and so on. If you're writing with a word-processor, be sure to find out whether the print-out is acceptable. If your paper is intended for a journal, obtain a copy of its requirements. Some journals carry a page of instructions to authors in each issue; others furnish one on request. For journals that don't do either of these, use a recent issue as a model. Much correspondence and rewriting can be avoided by preparing your paper for a specific publication and following its format faithfully. Practically all editors agree on one simple requirement: double-space *everything*!

At a few places in your report you will need to refer to yourself. The simplest way is to use the first-person pronoun: *I found Snarf's measurements to be accurate*. If your advisor or editor frowns on this usage, as some do, then simply say *the writer*--or, if you prefer to be more dignified, *the author*. Avoid use of the passive voice, as in *Snarf's measurements were found to be accurate*. This doesn't say who did the finding. If you did it, say so--in a straightforward, declarative sentence.

INTRODUCTION TO THE SENTENCE

The limestone yields fossils is a sentence, consisting of a subject (the limestone) and a predicate (verb plus object). Though a complete statement, it doesn't tell us much. To convey more information, we need modifiers. For example, the *gray thin-bedded* limestone *commonly* yields *silicified* fossils. Here we have added two adjectives to the limestone, an adverb to the verb, and an adjective to the fossils. With further decoration, we may write The gray thin-bedded limestone, *which is exposed in cliffs and low ledges,* commonly yields silicified fossils *from the residual soil.* This procedure adds a clause (a subsentence, with verb, introduced by the pronoun *which*) to the limestone, and a phrase (a sequence without verb, introduced by the preposition *from*) to the fossils. The *ledges* in the clause and the *soil* in the phrase have their own adjectives. The result is a much more informative sentence than the one we started with. Further elaboration is possible, but a point is soon reached where overloading may cause the framework to collapse. That would be the time to end the sentence and start a new one.

Seven of the eight parts of speech are contained in the sentence as finally written:

> *Nouns*: limestone, fossils, cliff, ledges, soil. These common nouns refer to physical objects, but other nouns are abstract (thickness, color), collective (team, genus), or proper (Dr. Snarf).
>
> *Pronoun*: which. This relative pronoun is both a stand-in for limestone and an introduction to the clause.
>
> *Verbs*: yields, is exposed. Another verb form, silicified, is used here as an adjective; it is the past participle of the verb silicify.
>
> *Adjectives*: gray, thin-bedded, silicified, low, residual. These words all modify nouns, which is what adjectives do. The little word *the* is an adjective form known as the definite article.
>
> *Adverb*: commonly. Like many adverbs, it is derived from an adjective by adding -ly. It obviously goes with the verb yields. Adverbs may also modify adjectives and other adverbs.
>
> *Prepositions*: in, from. Along with many other short service words--to, with, of, for, and the like--prepositions introduce phrases.
>
> *Conjunction*: and. Here it is the simple connective between nouns.

The eighth part of speech, missing from the sentence, is the interjection. This form expresses emotion. We have little or no use for interjections in science writing. Hurrah! This is the end of this section.

THE WANDERING MODIFIER

An obvious rule is that every modifier--adjective, adverb, phrase, clause--must be close to, indeed next to, the term that it modifies. Disregard of this rule results in unfortunate sentences. *Though quite poisonous to life, Dow believes these chemicals formed the first amino acids* seems to ascribe toxicity to Dow. It should read

Though these chemicals are quite poisonous to life, Dow believes they formed the first amino acids.

Phrases are especially likely to be misplaced. *Cumulative changes are shown as a function of time after the eruption in Figure 23* fairly cries out to have the final phrase moved to follow *are shown.* In *The acceleration of surface faulting is attributed to excessive use of ground water by some geologists*, the last phrase modifies *is attributed* and should follow it. In *About a dozen reversals have been discovered in the fossil record over the last 9 million years*, the phrase *over the last 9 million years* has nothing to do with the rate of discovery; it belongs after *reversals*, to which it clearly refers. *The Committee on Measurement of Geologic Time by Atomic Disintegration of the National Research Council* could well be referred to as *The National Research Council's Committee...* You are invited to reconstruct this sentence by simple phrase surgery: *Researchers recently investigated convection in a Newtonian fluid heated from within or below by numerical methods.*

Even clauses may be left dangling in the breeze. Consider the sentence *Many archeological discoveries have been made through the use of aerial photographs, some of which predate the Christian era.* The clause, with commas fore and aft, should obviously follow *discoveries. Production has been hampered by breaks in the pipeline, caused by landslides, which carry the oil to the coast*, records an unlikely function for landslides. We might recast the sentence *Production has been hampered by landslides causing breaks in the pipeline that carries oil to the coast.*

The verb form known as the participle--the present participle ending in -ing and the past in -ed--may introduce an independent phrase; and participial phrases have a reputation for dangling in mid-air. An absurdity like *Proceeding downstream, a fault was encountered* can best be corrected by leaving the phrase alone and indicating who did the encountering: *Proceeding downstream, we encountered a fault.* Another example: *Based on stratigrapic position, Orton regarded the shale as Devonian.* This seems to assign Orton a stratigraphic position. Better would be *Basing his conclusion on stratigraphic position, Orton regarded the shale as Devonian.*

A perennial problem is the floating adverb. In the sentence *Hopefully, the job will be done this week*, the adverb floats unattached to any other element in the sentence. It refers to the author, or rather to how the author feels--but this person is not present. Better to say *It is hoped the job will be done this week*; still better, *We hope the job will be done this week. Hopefully* is the main offender, but other floating adverbs have appeared in the geological literature. Examples include *Thankfully, the author has not done this* and *This study was gratefully supported by the National Science Foundation.*

Of course it is possible to use these adverbs correctly. I speak prayerfully when I say you won't misuse them. Just remember Mother Hubbard's dog: he followed her to the cupboard--hopefully.

THE VERB AND ITS DUTIES

In a sentence, the verb is where the action is. The active voice (the limestone yields fossils) generally produces more vigorous prose than the passive voice (Fossils are yielded by the limestone). This is certainly true when a simple or immediate action is involved. *I used the standard procedure* is short and clear; *The standard procedure was used* begs the question, By whom? (The passive voice is the favorite of bureaucrats: it tells what was done, but not who did it, thus avoiding assignment of responsibility.) But the decision rests on what is being emphasized. If you're writing about the Colorado River and its work, the natural way to state it is *The river has cut the Grand Canyon.* If your subject is the canyon, on the other hand, it's more reasonable to write *The canyon has been cut by the Colorado River.*

Here is a good sentence that leads off a paragraph on ocean surveying: *The deepest trench in the Atlantic Ocean was mapped recently.* In the next sentence we learn who did the mapping: *researchers using the GLORIA (Geological LOng Range Inclined Asdic) and its sea-floor scanning "fish".* To start the sentence with this mouth-filling expression, just to be able to put the verb into the active voice, would have been counterproductive. Nevertheless, when you find yourself writing *is measured, were determined,* and the like, stop and consider whether the active voice wouldn't do the job better. As noted on an earlier page, don't lean on the passive voice in preparing your abstract!

The verb must agree in number with its subject, even though the two may be some distance apart: a series of lava flows and tuff beds *is* next; evidence from field surveys and laboratory studies *suggests.* This rule is important in the use of nouns taken from the Greek or Latin, especially the common word *data.* Good writers know that this term is plural: *The data show this result.* An ill-informed writer states: *Little experimental data on low-gravity impacts exist.* If he considers data to be singular, then the verb should agree (*exists*). If he means data to be plural, the adjective should be *few* not *little.* This second alternative would be simple and grammatically sound. Or one could write *There are few experimental data on low-gravity impacts.* Either of these constructions would identify the writer as a person who knows a Latin plural when he sees one.

Such an expression as *2,000 feet of core* is considered to be singular, as the operative term is clearly the core, not the feet: *More than 2,000 feet of core was recovered.* Similarly, *About 100,000 tons of gravel is produced each year.*

Besides appearing in compound verb forms (*are checking, was tested*), the participle has several other uses. The -ing form, or present participle, may be used as a noun: *Understanding the processes is a goal of the program; Handling big problems on small computers is now possible.* Or it may stand alone: *Allowing for a small error, the figures seem reliable.* Such participial phrases must stay close to the term they modify. The participle may also act as a connective, as in a *probabilistic model using play analysis for petroleum exploration.* In this expression, *using* refers to *model*, but it also takes an object, *play analysis*, and so has elements of a verb. The past participle, ending in -ed or -en, is often used as an adjective: *silicified fossils, a broken specimen.*

The tense of a verb becomes important when you are citing the work of others. Choose one tense, preferably the past, and stick to it: *Although Snarf (1928) believed that the ore was deposited by magmatic fluids, Zilch (1985) found no evidence to that effect.* The fact that Snarf is long gone, whereas Zilch is young and active, doesn't justify using the past tense for one and the present tense for the other. Switching back and forth is a tricky operation and should be avoided.

No rule should be advanced to absolutely forbid splitting the infinitive--placing an adverb between *to* and its verb. However, a split infinitive is an awkward construction and is seldom if ever necessary. In the first sentence of this paragraph, the word *absolutely* could well be omitted. When an adverb must be retained, place it before, or more commonly after, the infinitive. *It is essential to carefully calibrate the instrument* should read *It is essential to calibrate the instrument carefully.* Or you could avoid the infinitive by writing *The instrument must be carefully calibrated.* It seems unnecessary to further emphasize the point--or rather, to emphasize the point further.

NOUNINESS: ITS CAUSE AND CURE

Geophysicists use the term *noise* for any undesired sound or disturbance within a given frequency band; it is also referred to as *rock noise.* Now if you were writing a report on locating the source of rock noise, wouldn't it be logical to call it just that? It would--but that isn't the way the report came out. It was titled *Rock noise source location techniques*--five nouns in a row. This is an example of noununess, or nounspeak, a usage that has become epidemic, especially in scientific and technical writing. We have already seen several examples in the section on preparation of a title. Editors want material condensed in order to save space and money, and many authors are happy to oblige, simply by piling up nouns, without hyphens, prepositions, or other connectives. Five nouns, as in the example above, by no means set a record. Here is a nine-nouner: *Murgul mine development and concentrator debottlenecking investment, expansion and modernisation project.* The problem with effusions like that is that readers must plow through a long string of nouns before getting to the subject, and then work their way back down the string to find out what is said about it.

The situation gets further out of hand when adjectives are mixed in with the nouns, forming a sort of clotted prose. In the expression *incoherent scatter radar user community*, for example, it is not at once apparent that it's the *radar*, not the *user community* (=users) that is incoherent. *Users of incoherent scatter radar* is an improvement. The governor of California can, in case of need, call on a *seven-person mudslide advisory panel*--prompting derisive questions about seven-person mudslides. As a bit of practice, try recasting this photo caption that appeared in an internationally read journal, below a picture of what looked like a 55-gallon steel drum: *Liquid metal fast breeder reactor spent fuel shipping cask.*

Beware, then, of stringing nouns together, with or without adjectives. Think, and write, using the prepositional phrase: the phrase is the answer to acute nouniness. *Stack emissions desulfurization sludge disposal* is a reader-baffling expression that need never have happened. *Disposal of sludge from desulfurization of stack emissions* says it in English and serves the reader.

A simpler form of nouniness is the three-word expression consisting of an adjective followed by two nouns. In this combination, which I term a lulu, the reader is not sure which noun the adjective goes with. Examples appear in common speech all the time. We can get insurance against scheduled airline accidents; we may meet a small college professor; at the supermarket we deal with a frozen food clerk. In this common variety of lulu, the adjective modifies the first noun--as it also does in such expressions as high level terrace, crude data analysis, and 38th parallel lineament--*but this fact is not instantly apparent to the reader.* Certain high terraces are level, certain analyses are definitely crude, and maybe there are a whole lot of parallel lineaments, of which this is the 38th. The simple and obvious cure for the lulu is the hyphen, which makes the first two words a unit modifier of the third: high-level terrace, crude-data analysis, 38th-parallel lineament. Same with mean-dip map, low-angle fault, precious-metals staff, and so on. Use of the hyphen conveys instant clarity. Inserting it is analogous to focusing a slide on a screen.

A word of warning: some editors dislike hyphens and will take them out of your copy if they can get away with it. Their argument is that the first two words commonly go together and therefore it isn't necessary to hyphenate them. Nonsense. *Crude data analysis* is crude writing. Your job is to see that your reader doesn't stumble, even momentarily. If *rare* and *earth* go together, link them. Don't let an editor dehyphenate your unit modifiers!

It is a pleasure to vary the monotony by quoting a good example. Note the use of unit modifiers and phrases in this expression, which is a model of clarity: *Finite-element two-layer model for simulation of ground-water flow.*

PRONOUNS TO THE RESCUE

A pronoun is a word used in place of a noun. It mentions a thing without naming it. The noun to which the pronoun refers is its antecedent.

In technical writing, the little impersonal pronoun *it* is very useful. In the second sentence above, there are two *its*, each with an antecedent: *It* (the pronoun) *mentions a thing without naming it* (the thing). The possessive of *it* is *its*, with no apostrophe. Don't confuse *its* with *it's*, which stands for *it is*: *It's a locality famous for its minerals.*

Which is used to introduce a nonrestrictive or commenting clause; such a clause does not limit or define, but merely adds something. It must be set off by commas, as in these examples: The formation, *which is famous for its fossils,* is of Jurassic age; The crushed glass, *which contains titanium,* did not melt; Many of the earthquakes occurred in swarms, *which were associated with strong local uplift.* In each of these examples, the statement introduced by *which* is an extra, parenthetical bit of information. The sentence makes sense without it.

That, in contrast, introduces a clause that is essential to the meaning of the sentence. It is not set off by commas. *The information that I need is in his book. Try to avoid the pace that kills.* The information, and the pace, are both defined by the *that* clause. *The mouse that roared* is a particular mouse; *the rocks that form the cliff* are those rocks and no others. It is not a misdemeanor to introduce a restrictive clause with *which* instead of *that;* but somehow *This is the house which Jack built* doesn't sound right. The converse is not advisable, as *that* is always limiting or restrictive. *The California Academy of Sciences (that houses the largest Galapagos collection in the world)* is incorrect, because it implies that there is another California Academy of Sciences that does *not* house that collection.

A main purpose of *it, which, this,* and *that* is to help in avoiding needless repetition. Many beginning writers, having got hold of a resounding term or phrase, tend to repeat it indefinitely, with resulting stupefaction on the part of the reader:

> This report deals with the geology of the Dull Thud mining district. The Dull Thud mining district is located in Abysmal County, Nevada. Several geologists have mentioned the Dull Thud mining district, but no one has described the mining district in detail.

Note how the use of three pronouns shortens this and converts it into readable prose:

> This report deals with the geology of the Dull Thud mining district, *which* is located in Abysmal County, Nevada. Several geologists have mentioned *this* district, but no one has described *it* in detail.

The two sentences that follow are from the first draft of a PhD dissertation. Can you convert them into readable English, in large part by the use of pronouns? What was the person trying to say, anyway?

> The authors argued that the biotite occurring in the tuff is euhedral and does not appear to be weathered as much as the biotite in the Carboniferous granite. However, it is possible that the biotite in the tuff appears fresher than the biotite in the granite because the biotite in the tuff has weathered less relative to the biotite in the granite after its deposition in the tuff.

Here are three further recommendations. 1) Be sure that the antecedent of each pronoun is clear. In *The color of the cementing material in the sandstone, which was quite distinctive,* we don't at once know whether it was the color, the cementing material, or the sandstone that was distinctive. *The distinctive color of the cementing material in the sandstone* would be an improvement. 2) Don't use *which* where its antecedent is a whole clause, as in *Some of the rocks contain no fossils, which makes correlation difficult*. There are better ways to say it, one of which is *Correlation is difficult, because some of the rocks contain no fossils*. 3) The pronoun *whose*, which generally refers to a person or persons, may also be used for inanimate objects, as in *a wide range of rock types, whose ages are well bracketed. Whose ages* is shorter and less stilted than its alternative, *the ages of which*.

STARTING AND STOPPING

Starting sentences with *There is, It is*, or *There are* should in general be avoided, on the ground that those are just extraneous words. *There is an offset of 50 feet on the fault* means the fault has an offset of 50 feet. *There are many fossils in the limestone* could be better stated *The limestone contains many fossils*. But this is not a flat prohibition. It would be hard to find a better form of expression than *There is no accounting for tastes*, or even *There is a tavern in the town*. N.H. Darton, who mapped and reported on much of the American Southwest for the U.S. Geological Survey, started a description of "The General Character of the Rocks" by writing *There are in New Mexico many kinds of metamorphic, sedimentary, and igneous rocks*. This can hardly be improved by rephrasing. But we are not all N.H. Dartons. If you detect an excess of *there is* and *there are* in your writing, better reconsider.

Stopping when you're through is also advisable. For years I commuted on rail coaches that carried this stenciled message: *Passengers must not put head or arms out of the windows. Serious injury or loss of life may result from neglect of this notice*. The anonymous author should have stopped after *may result*. The remaining five words are not only unnecessary; they falsify the whole message. For thousands of commuters neglected the message every day and none was seriously injured or lost his life as a result. *The Jacksonburg seas deposited limestones which were at first pure but later argillaceous limestones were deposited* would be immensely improved by knocking off the last three words (and putting a comma before *which*). Stopping when you're through is one aspect of a famous dictum put forth many years ago by Professor William Strunk of Cornell University: omit needless words! That short sentence deserves your close attention.

AT EASE AMONG THE COMMAS

The comma marks a slight pause in the flow of words, for emphasis, change of pace, or change of direction. If you have a good ear for speech--for how a sentence sounds--you should have no trouble. To decide whether a sentence needs one or more commas, read it aloud.

The comma separates the items in a simple series, as in *programs involving hydrology, geology, geochemistry, rock mechanics, and geostatistics.* Note the comma before the *and.* There is no stern rule that calls for such a comma, and it is often omitted. Occasionally, such omission may cause confusion or embarrassment, as when a geological publication gave its subscription rates for members "in the United States and its possessions, Canada and Mexico." Canadian and Mexican subscribers were not amused. Your editor may have strong feelings about the pre-*and* comma; if so, humor him unless misunderstanding might result. In any event, be consistent.

The comma has several other workaday uses. It separates adjectives preceding a noun when they refer to the same or similar properties: *calcareous, siliceous, and phosphatic sediments*; *in an organized, logical way.* Your ear tells you that certain transitional or explanatory terms must be enclosed by commas: *Today, however, it is different*; or *It was better, therefore, to change the procedure.* Dates and place names, of course, are set off by commas: *The meeting will be in Orlando, Florida, on January 10-13, 1990.*

As we saw earlier, the nonrestrictive clause, which is not essential but merely comments, must have commas fore and aft--as it does in this sentence. This same rule applies to the rarer *who* clause. Omitting the commas had an unfortunate effect on this sentence about seismic prospecting: *Then comes the shooter who detonates the charge and the recording vehicle.*

Commas do not belong in a series of adjectives that each refer to a different property. *Intensive shallow seismic profiling* is written without commas, as is *a 700-watt 2.45-gigahertz microwave oven.* Also, you should be aware of a few place names that look as though they should have commas but don't. Baja California is a peninsula, Trans-Pecos Texas is a region, and Papua New Guinea is an Australian territory. Write them without commas!

Omission of commas may result in what Vanserg (1952) calls the how-the-hell-did-we-get-here sentence. It seems to be going along fine when suddenly it blows up. Quoting Vanserg:

> "Geological sections on the footwall and hanging wall show correlation of various rocks and vertical movement of 1400 feet. . ." That seems to be straightforward enough, but wait! You haven't come to the end of the sentence. It goes on: "is reached." See? You have to come back and start over, not missing the dog-leg at the second "and."

Here is another fine example: *The estimator knew that in drilling the rock-shattering mechanism is partly shear of the brittle materials.* By the time the reader has figured out that no one was drilling the rock-shattering mechanism, and has gone back and mentally inserted the necessary commas around *in drilling*, he is likely to have forgotten the original subject and to have lost interest in the whole project.

The author of this sentence was a nonbeliever in commas. Read it out loud and find two places where a comma is needed: *In general end points are defined from statistical analysis of crossplots and linear equations are used to quantify values.*

Many people have strange ideas about use of the comma--for example, "When in doubt, leave it out," or "It's impossible to have too many commas." Ignore such nonsense. Remember the purpose: if a slight pause in the flow of words is indicated, insert a comma; otherwise, no. This advice has been partly ignored in the following expression, which makes no sense. Read it aloud and restore its meaning by inserting a second comma: *long-term disposal of nuclear waste and spent fuel, from commercial reactors in various geological formations.*

HOW TO QUOTE

When you wish to incorporate someone else's words into your text, enclose them in quotation marks, or quotes: generally single (' ') if English, double (" ") if American. When you use quotes, you must be sure that the quoted words are copied exactly as written by the original author: *Jahns (1955) terms pegmatites "both the delight and despair of persons attempting to work them for materials or information."* You may, of course, rephrase Jahns' statement in your own words--as, *Jahns (1955) remarks that attempting to work pegmatites leads to both delight and despair.* This procedure requires no quotes, but it does require attribution. To write the comment about pegmatites as your own, without naming the author, would be unethical.

Suppose you come across a sentence of which you wish to quote only a part. You may omit any words you wish, but each omission must be indicated, commonly by three periods spaced apart. Let us say that the original sentence in full, from Snarf (1978), reads: *Recent work has shown that the magnesite cuts across the bedding of the dolomite, preserving some of its structures, and is of replacement origin.* You may elide the sentence thus: *According to Snarf (1978), "recent work has shown that the magnesite . . . is of replacement origin."* The quoted words must be verbatim.

Long quotations--a paragraph or more--may be set off from the rest of the text by being typed in narrower format--i.e., with wider margins (but still double-spaced!). In such situations, quotation marks are not used. The other requirements (attribution, verbatim copying, elision) still hold.

Another use of quotation marks is to set off informal, slang, or other out-of-the-ordinary terms: the Black Hand Sandstone, or "Big Injun Sand"; giant crystals, or "logs", of beryl.

OTHER POINTS ON PUNCTUATION

The **semicolon** serves as a less-than-full stop between related parts of a sentence. *Many are called; few are chosen. The chief requirement is uniformity; a lack of uniformity rules out much of the stone.* This is a handy usage, but it is easy to overdo. The semicolon is also used between items in a series when these items are long or contain internal punctuation; in other words, when the comma won't do the job:

> The section includes, in ascending order, a basal quartz-pebble conglomerate, 32 feet thick, which is highly cross-bedded; shaly sandstone, 40 feet thick, with persistent silty layers; and a thin shale with abundant fossils.

Don't use the semicolon as an introductory mark. Introductions are the job of the colon. Never write *Dear Sir;* or *As follows;*.

The **colon**, as you have probably noted from these paragraphs, is the mark that introduces, or tells what's coming. Although it's useful, in shorter expressions it can often be dispensed with. All three of these are acceptable, but the best one is the last:

> The most abundant elements are as follows: oxygen, silicon . . .
>
> The most abundant elements are: oxygen, silicon . . .
>
> The most abundant elements are oxygen, silicon . . .

The **dash** may occasionally be used instead of a conjunction or a semicolon, to separate balancing clauses: *no vestige of a beginning--no prospect of an end.* Or a pair can be used to set off terms for emphasis: *Two types--veins and replacement bodies--are of value.* Or the dash may lead into an illustrative term: *Gypsum may be imported--for example, from Nova Scotia.* Don't confuse the dash with the hyphen. Dashes separate. Hyphens connect.

The **hyphen** joins unit modifiers, as mentioned earlier. Use it whenever two words, or even three, modify a following word: 20th-century research, second-order effect, a two-mile-long pegmatite. Don't place a hyphen after adverbs ending in -ly: steeply dipping beds, recently added equipment. If you're using a word processor, check your copy for words that are broken at the end of a line and continued on the next line. Machines are programmed to drop in a hyphen, if needed, after such syllables as *be-*, as in be-tween; or before *-th*, as in ar-thropod. How, then, about be-drock or an-thill? Your eye must be alert for items like these on processed copy or proof sheets.

The **apostrophe** is the mark that denotes possession. It comes before the *s* if the possessor is singular, and after the *s* if the possessor is plural: *the sea's depth; the rocks' age*. "The study area on the Vier's farm" is wrong, even though it appears in a paper by a former president of a prestigious geological society. If the farmer's name is Vier, it should be *the Vier farm* or *the Viers' farm*; if his name is Viers, either the *the Viers farm* or *the Vierses' farm*. The gentleman who developed the scale of hardness that most of us use was named Mohs. *The Mohs hardness scale* is fine. If you insist on using an apostrophe, put it outside, not inside, the name. The possessive pronoun *its*, as in *its cleavage is cubic*, has no apostrophe, and there is no apostrophe in plurals, geological or other. An exception to the last statement may be found in such terms as *the 1960's*, but *the 1960s* will do the job as well if not better.

The **diagonal or oblique stroke** is useful in connecting related expressions that the hyphen cannot readily handle. Reference to the *Anthracite-Crested Butte area* of Colorado calls up a picture of a butte with coal on top. (Actually, Anthracite and Crested Butte are communities.) The confusion is avoided by writing *Anthracite/Crested Butte*. A *rare earth-thorium association* raises the questions, what is an earth-thorium association, and why is it rare? Rewrite it *rare-earth/thorium association* (hyphenating the unit modifier). Similarly with expressions like *Big Salmon River/Goose River area* and *Golden/Green Mountain district*. The expression and/or, as in *The data may be shown in tables and/or graphs*, does not offend the writer of this book, but many editors don't like it. You could write *The data may be shown in tables, graphs, or both.*

Parentheses have obvious uses in such expressions as (Snarf, 1958) and $(SiO_4)_3$. They may also set off an explanatory phrase or comment, as in *Small amounts of ulvospinel (a refractory iron-titanium oxide introduced incidentally with the ilmenite) did not completely melt.* In such sentences, parentheses are not required but are merely an option. Parenthetical material may be indicated just as well by dashes, or simply by commas.

AMERICAN ENGLISH AND BRITISH ENGLISH

A few geological features or materials have entirely different words in these two "languages"; for example, diabase in America is dolerite in Britain. But many words differ only in their spelling. Some of the more common ones follow.

American Usage	British Usage
aluminum	aluminium
amygdule	amygdale
analyze	analyse
catalog	catalogue
center	centre
color	colour
dike	dyke
draft	draught
eolian	aeolian
fiber	fibre
gray	grey
hematite	haematite
harbor	harbour
meter	metre
organize	organise
orient	orientate
paleontology	palaeontology
Paleozoic	Palaeozoic
program	programme
recrystallization	recrystallisation

Occasionally, different usages in the two countries produce anomalous results. For example, an American textbook in sedimentology appeared in its British edition with the subtitle *How Strata Get Laid.* Translation, even from American to British English, is always a risky business.

CAPITALIZATION

It is not possible to lay down hard-and-fast rules on capitalizing, as preferences differ with editors and publishers. Some general recommendations follow. Formal geologic names, geographic names, and combinations of the two should be capitalized:

Jurassic System
Oxfordian Stage
Morrison Formation
Old Red Sandstone
Atlantic Coastal Plain
Potomac and James Rivers

North Downs
Strait of Dover
Cincinnati Arch
East Texas Coal Field
North Sea Basin
Williams Range Fault

As to the names of persons, one inflexible rule is: spell it right! Use whatever combination of given names and initials the individuals use themselves, unless the journal for which you are writing has a standard form to be followed (e.g. initials only).

Don't capitalize once-proper names that are no longer identified with the persons or places from which they were derived:

canada balsam	nicol prism
carlsbad twins	plaster of paris
diesel engine	portland cement

Capitalize phylum, class, order, family, and genus, but not species.

EXOTICS AND THEIR PLURALS

Several nouns in common use in science are taken directly from the Greek or Latin. You should know their singular and plural forms and use them correctly.

Type	Singular	Plural
the *-on* nouns	criterion	criteria
	phenomenon	phenomena
the *-a* nouns	costa	costae
	formula	formulae (or formulas)
	sicula	siculae
	vita	vitae
the *-us* nouns	focus	foci
	fungus	fungi
	locus	loci
	modulus	moduli
the *-um* nouns	bacterium	bacteria
	curriculum	curricula
	erratum	errata
	medium	media
	memorandum	memoranda
	phylum	phyla
	quantum	quanta
	spectrum	spectra
	stratum	strata
	symposium	symposia
the *-ies* nouns	facies	facies
	series	series
	species	species

Missing from the above list of *-um* nouns is *datum*, with its plural, *data*. These terms don't have the conventional relationship. Datum means a base line or reference point. A manufactured plural, *datums*, is in use by seismic stratigraphers among others. Data, of course, refers to information, generally quantitative. It is a plural form: *These data are accurate*. If you're embarrassed to write *Many data are available*, write *Much information is available*. Use of *data* as a singular brands a writer as only semi-literate.

LATIN SHORTHAND

From various sources, including classics and the law, we have inherited a group of abbreviations in Latin. The four listed below are occasionally handy, although each has a perfectly good English equivalent.

e.g. (*exempli gratia*) means "for example." It should be preceded by a comma: Correlation may be made on microfossils, e.g. conodonts.

et al. (*et alii*) means "and others." As *et* is a complete Latin word, there is no period after it. The expression is generally used in citations, within the text, of papers with more than two authors, e.g. (Whillams, J.S., et al. 1955).

etc. (*et cetera*) literally means "and other things." In practice, it means "and so forth." It is a good term to avoid, as you can get into the habit of using it when you've run out of anything to say, or don't want to bother to be specific: The rock consists of feldspar, quartz, etc.

i.e. (*id est*) means "that is" or "namely." Put a comma before it, as with e.g., but don't confuse it with e.g. His work involves a group of microfossils, i.e. conodonts.

NUMERALS VS. WORDS

A good general rule is to use numerals for all units of measurement (feet, percent, dollars, years, and so on), and for all other items of 10 or more. Write out the number in words only for non-measurement items of less than 10: four quarries, nine readings. If a quantity is less than 1, put a zero in front and use the singular: 0.8 mile, 0.75 gallon. (But 1.01 gallons.)

Most publishers have their own rules. Differing preferences are nicely illustrated in the matter of punctuating numerals. According to the "house style" of one publisher, neither 4000 nor 50000 has a comma; but the U.S. Geological Survey, following the Government Printing Office, uses 4,000 and 50,000. It's wise to follow the usage prescribed by your publisher, or at least to have an authority to refer to in case of argument. In any event, be consistent.

PARALLEL CONSTRUCTION

The coordinate conjunctions--but, and, or, and a few others--require the same construction on either side. In the sentence *The living group appears to be related to the angiosperms, but its fossil record is virtually nonexistent*, a statement about the living group is balanced by one about its fossil record, the fulcrum being the conjunction *but*. Similar parallelism appears in *the early history and subsequent evolution of Mars*, and in the sentence *Pure hematite did not respond, or couple, to the microwave radiation.* In *seismic and bathymetric studies and dredge hauls*, the first *and* connects adjectives and the second connects nouns. The expression is technically correct, though not especially graceful.

The sentence *The shale was thick, fissile, and contained no fossils* is faulty because the *and* connects adjectives with a verb. It should be rewritten *The shale was thick, fissile, and unfossiliferous.* Items in a list or series should always be of the same type.

Parallel construction is necessary not only with single conjunctions but also with correlative pairs. An example (*not only/but also*) appears in the sentence you just read. Other examples: *Neither the chemistry nor the plant inclusions proves the botanical origin of the amber. Both geophysical and geochemical constraints apply.*

Here are a couple of sentences from a progress report on petrology: *Researchers have discovered that at depths of 100 km or more, olivine phenocrysts float in the magmas from which they crystallize. Consider the implications of that discovery for the process of igneous differentiation: in deep magma chambers olivine floats but garnet sinks.* The last five words, a neat parallel construction, leave a vivid impression on the reader's mind.

PLAIN AND FANCY

Will you write *about* or *approximately*? *clayey* or *argillaceous*? In English we have many twins and a few triplets in which the terms come from different languages but mean essentially the same thing. The shorter form is Anglo-Saxon, the longer one is derived from Latin. Which one you use will depend perhaps on how impressive you wish to sound. Keep in mind, however, that the good writer tends toward the shorter words.

about, approximately	many, numerous
clayey, argillaceous	sandy, arenaceous
find, locate	show, exhibit
form, constitute (both Latin)	start, begin; initiate

REDUNDANCY

The temptation to emphasize an already emphatic term, by introducing a modifier, is easy to yield to. But most such effort is misdirected, as the added word is not needed. In every example below, the term in the right-hand column says what needs to be said better than its matching term in the left-hand column.

red in color	red
entirely absent	absent
totally destroyed	destroyed
partly damaged	damaged
age dating	dating
most unique	unique
really significant	significant
completely unnecessary	unnecessary
actual photographs	photographs
widely surrounding	surrounding
low-lying depression	depression
mineralized ore	ore
alternating interbeds	interbeds

If there were an award for redundancy, it would doubtless go to the author who wrote *The chief ore control extends throughout most of the entire region.*

TIME LIMIT

Don't use words that designate time when you mean something else. In each of the following sentences, the term in parentheses should be substituted for the one that precedes it.

The unit is sometimes (locally) absent.

The shales are often (commonly) crumpled.

When (Where) the sandstone crops out it forms a cliff.

The fossils are usually (generally) well preserved.

The sentence *Salt was deposited in the Late Silurian while shale accumulated in the Early Devonian* seems to imply an impossible time warp. Replacing *while* with *whereas* would clear it up.

VERBING

Making verbs from other parts of speech, chiefly nouns, has been going on for a long time and is a recognized way of expanding the language. If a meeting votes to table a motion, no one is concerned about the use of *table* as a verb. Sales representatives *contact* purchasing agents; editors *critique* manuscripts. The earth sciences have their share of new coinages. Hematite *pseudomorphs* pyrite; certain rocks *source* oil; digital terrain data are *mosaicked.* A common practice is to add *-ize* (or *-ise*) to nouns, as in saying that an ore body is heavily *stringerized.* (Why not? We say *mineralized* without hesitation.) Many such verbs are useful and may remain in the language. Others are marginal, to say the least. It is doubtful that *total depth* will become an accepted verb, as in *The well was total depthed at 7,000 feet.* The same may be said of the verb in this sentence: *It won't cash flow at $80 an acre.*

There is no law against converting a noun into a verb; anyone may do it. If you are tempted, however, consider carefully whether the action is needed. The recommendation here is to be a judicious follower, not a bold leader, in the verb-coining business.

Occasionally a verb may be invented, not adapted from another part of speech. Geophysicists surveying the ocean floor use the recently minted verb *insonify* "to describe the action of permeating a volume or covering a surface with sound waves. It is analogous to the word *illuminate* to describe the action of permeating a volume or covering a surface with electromagnetic waves in the visible range." In this instance a clear need existed and a descriptive verb was coined from scratch.

TRICKY TERMS

age, date

You date the rock; the result is the age. Avoid "age dating."

altitude, elevation, height

Airplanes fly at altitudes; maps give elevations; hills stand at a height above the surrounding terrain.

bearing, containing

Compare these sentences: *The shale is overlain by limestone bearing fossils* and *The schist is intruded by mica bearing pegmatite.* The first sentence is OK; the limestone bears the fossils. But the second is not, as the mica does not bear the pegmatite. Why not insert that hyphen right now, to make mica-bearing a unit modifier? *Cobalt containing impurities* means an impure cobalt; *cobalt-containing residues* refers to certain mill wastes.

between, among

Use *between* for two objects, *among* for three or more: distance between the two ridges, differences in elevation among the five peaks.

case, instance

These are good terms to avoid, as they really don't say anything. *In one case the section is well exposed* could be better stated *At one place the section is well exposed. In many cases the records have been poorly kept* means *Many of the records have been poorly kept. This is the only instance in the area of an underground limestone mine* simply says *This is the only underground limestone mine in the area.*

comprise

Comprise means to include or contain: *The formation comprises three facies.* The reverse relation requires *form* or *constitute*: *Three facies constitute the formation.* Avoid the passive form, *is comprised of.* In fact, avoid *comprise* altogether. There are so many good alternatives that you don't need it: include, contain, be made up of, consist of.

conditions, purposes

These are vague, general words that you can commonly get along without. *Under proper drainage conditions the land could be made suitable for farming purposes* just means *With proper drainage the land could be made suitable for farming.* The last word of this sentence should be deleted: *The stone is used for building purposes.*

develop

A stream may develop meanders, or a mining company may develop a deposit, but rocks ordinarily do not develop. *Thick sandstones are developed farther to the west* is written to mean *Thick sandstones occur farther to the west.*

due to, owing to

Use *due to* if you mean caused by: *The flood was due to rapid snowmelt.* Use owing to if you mean because of: *The streams flooded owing to rapid snowmelt.*

Early, Lower

Early refers to time, *Lower* to rocks. Lower Permian rocks were deposited in Early Permian time.

farther, further

Farther is literal and measurable: *farther up the valley. Further* is figurative: *further study; on further investigation.*

fewer, less

Fewer refers to separate items or objects: *fewer outcrops. Less* refers to a single term: *less precipitation.*

horizon, zone

If you are a soil scientist, *horizon* means a layer of soil. But to the geologist, a horizon is a surface, with no thickness: *The top of the Trenton Limestone is the horizon contoured.* Oil and ore don't come from horizons, no matter how often you see the terms used that way. A zone, on the other hand, is a unit of rock: *The lower Trenton Limestone is the producing zone*, or *Fluorite occurs in the ore zone.* Biostratigraphers and mineralogists use *zone* in specialized senses.

important

This is a dignified term that sounds, well, important. If used, it should be modified: *These rocks are commercially important.* Often you can substitute a more specific term. *These minerals, named in order of their importance* probably means

These minerals, named in order of their abundance. The sandstone is an important ridge-maker might be better expressed *The sandstone makes pronounced ridges.*

Late, Upper

Late refers to time, *Upper* to rocks. Upper Permian rocks were deposited in Late Permian time.

lay, lie

Lay (laid, laid) is a transitive verb: it must have an object. *I now lay the thin section on the same shelf where I laid one yesterday. Strong winds laid down many feet of loess. These are water-laid silts.*

Lie (lay, lain) is an intransitive verb: it takes no object. *Pebbles lie on the beach. In the Silurian, salt lay on the shale.*

lithology

Lithology is the study of rocks, not the rocks themselves. *The formation consists of several different lithologies* should be written *The formation consists of several different kinds of rock.*

majority

This term must be followed by a plural: *the majority of the voters, the majority of the pebbles*. If you're writing about a singular subject, use *most*: *most of the ore, most of the ice.*

Middle

This term refers to an interval of geologic time between Early and Late, and also to the rocks formed during that interval. Middle Jurassic granites were intruded in Middle Jurassic time.

offshore

This term, much used by petroleum geologists, has long been an adjective (*offshore production*) and an adverb (*a well located offshore*). Today it is also a preposition: *The test well is located offshore Western Australia.*

over, more than

Over is an informal term, common in trade journals: *Production was over 500,000 tons.* In serious writing, use *more than.*

overlie, underlie

These verbs are, in effect, transitive, because they mean "lie over" and "lie under." *Shale overlies the sandstone. In the Silurian, salt overlay the shale. These rocks are underlain by granite and overlain by residual soil.* There is no such term as *overlaid* or *underlaid.*

production

The production of a mine or an oil field is a statistic, expressed in tons or barrels. It is not a physical entity. In the sentence *Most of the production is shipped by rail*, the word *product* could be used; better would be oil, ore, kaolin, coal, or whatever is being discussed.

quite, very

Neither of these adverbs does much for your writing. Rather than saying *The formation is quite thick* or *The phenocrysts are quite abundant*, state how thick or how abundant (if you don't know precisely, give a range or say *about*). *The water is very deep* begs for more precision. A very steep cliff is just a steep cliff.

suffering and responsibility

The student who wrote *The Appalachians suffered a deeply moving erosional experience* was imparting feelings to the mountains that they do not possess. To say *The mine enjoys an excellent location* suggests that the mine has expressed an opinion. Similarly with responsibility. *A submarine fault was responsible for the tsunami* just means the fault *caused* the tsunami. A good rule is to avoid endowing geologic features or processes with human traits.

type

Type is a perfectly good word, with various meanings, but it lends itself to a locution that you ought to avoid. *Channel type sands* are just channel sands; *Lake Superior type iron ores* could better be expressed as iron ores of the Lake Superior type. A common expression beloved of economic geologists,
Mississippi Valley type ore deposits, is poor type prose.

unique and other absolutes

It is generally agreed that unique, like pregnant and dead, is an absolute: it cannot be more or less. Yet in the literature we see expressions like *unique enough to be of broad application*. Another absolute was abused by a distinguished lecturer who told his audiences that, owing to the discovery of many ore deposits, the earth has become *less virgin*. An author wrote that pyrite may be *unusually ubiquitous* in the Wind River Formation. Obviously it is best to allow unique and similar terms to stand alone.

SPELL IT RIGHT!

Here is a short list of words that seem to give geologists trouble. It will be worth your while to get the correct spellings firmly in mind.

asymmetrical
consistent
desiccate
eustasy
fluorite
fluorspar
isostasy
liquefy
liquid
occur, occurred, occurrence
permeable, permeability
persistent
phosphorus
predominant
resistant
soluble

WATCHEMS

Different words that are similar in sound or spelling, and thus may be confused, are termed *watchems* in newspaper offices. Listed below are several pairs of such words, and one triplet, with brief definitions. The definitions are far from exhaustive; they are simply intended to make it easy for you to differentiate between similar words. Some sample sentences, tilted toward earth science, are included. If you need more information, reach for your desk dictionary. Better to be fully informed than embarrassed.

absorption, adsorption

Absorption is a general term meaning assimilation or incorporation. (Soils may become unstable owing to absorption of water.) *Adsorption* is the adhesion of ions or molecules of a gas or liquid to the surfaces of solid bodies with which they are in contact. (Certain clays remove coloring matter from oils and waxes by adsorption.)

aerial, areal

Aerial means pertaining to the air, as in aerial photography. *Areal* means pertaining to an area, as in areal reconnaissance. Since both terms may be applied to mapping, surveying, and the like, it is important to distinguish between them.

affect, effect

Affect: to influence, or have an effect on. (Climate affects weathering.) *Effect*: to bring about, produce, or accomplish. (Constant rains effect solution of the rock.) *Effect* may also be a noun, meaning anything brought about by a cause or agent; result. (Climate has an effect on weathering.)

alternate, alternative

Alternate: by turns; one after another. (We found alternate exposures and covered intervals.) *Alternative*: being or providing a choice between two things. (We laid out alternative routes.) *Alternate* may also be a verb, meaning to switch back and forth, and *alternative* may be a noun, meaning choice.

cite, sight, site

Cite: to quote or give as an example. (I often cite the work of Snarf.) *Sight*: to see or observe, especially through a surveying instrument; also, anything seen. (We sighted the triangulation flag; it was a welcome sight.) *Site*: to place or locate; also, a place or location. (They sited the reactor near a fault line. This site was considered unsuitable.)

coarse, course

Coarse: consisting of large particles. *Course*: progress, channel, direction, and numerous other meanings.

complement, compliment

Complement: to complete or fill out. (This research complements his earlier work.) *Compliment*: to express admiration, courtesy, or respect. (I compliment you on your writing.) Both words may also be used as nouns.

desert, dessert

Desert: a dry, barren region, e.g. the Sahara or Antarctica. *Dessert*: pie a la mode.

envelop, envelope

Envelop: to surround or cover completely. (An alteration product envelops the original mineral.) *Envelope*: a thing that envelops; a wrapper or covering. (The pyrite was surrounded by an envelope of limonite.)

foreword, forward

Foreword: an introduction, as to a book. *Forward*: at or toward the front; advancing.

imply, infer

Imply: to hint, suggest, or intimate. (The rock's texture implies slow cooling.) *Infer*: to conclude or decide from something known or assumed. (We infer that the rocks are of deep-water origin.)

inter-, intra-

Inter- means between or among; *intra-*, within or inside of.

intercept, intersect

Intercept: to stop or interrupt the course of; cut off. (We intercepted an SOS.) Intersect: to cross or cut across. (The well intersected a fault.)

its, it's

Its is the possessive of the pronoun it; *it's* is a contraction of *it is*. (It's a locality well known for its minerals.)

lead, led

Lead is the present tense of the verb meaning to direct, guide, or conduct; *led* is the past tense.

perspective, prospective

Perspective: the showing of an object in a drawing as it appears to the eye with respect to relative distance or depth. (The diagram showed a fault block in perspective.) *Prospective*: looking toward the future; expected; likely. (They talked of a prospective mining venture.)

precede, proceed

Precede: to go before in time, place or importance. (Heavy rains preceded the landslide.) *Proceed*: to go forward, or carry on some action. (We proceeded to climb the mountain.)

principal, principle

Principal: first in rank; chief or main. (Orthoclase is the principal mineral.) Also, a governing officer, as of a school. *Principle*: a fundamental truth, law, or doctrine. (He cited the principle of differential entrapment of petroleum.)

remanant, remnant

Remanant: said of the component of a rock's magnetization that has a fixed direction relative to the rock and is independent of moderate applied magnetic fields. (The basalt showed strong remanant magnetization.) *Remnant*: what is left over: remainder or residue. (Erosion had left only a remnant of the sandstone.)

shear, sheer

Shear: a deformation resulting from stresses that cause adjacent parts of a body to slide past each other in a direction parallel to their plane of contact. (Faulting is an expression of shear in the crust.) *Sheer*: perpendicular or very steep, as the face of a cliff. (It was a sheer drop of 80 feet.)

simulate, stimulate

Simulate: to look or act like. (A surface of graphic granite simulates cuneiform writing.) *Stimulate*: to rouse, excite, or invigorate. (Damage from the landslide stimulated heroic rescue efforts.)

terrain, terrane

Terrain: a region of the earth's surface considered as a physical feature, an ecological environment, or a site of some planned activity. (The invasion was preceded by aerial analysis of the terrain.) *Terrane*: a general term applied to a group of rocks and the area in which they crop out. (The ultramafic terrane is thought to be of oceanic origin.)

BE YOUR OWN EDITOR

On an early page of this book you were urged to plan before writing: outline first, then write. To this recommendation we can now add another: cool off, and revise. After you have completed a section of your report, set it aside for a few days. Then pick it up and read it as though it were someone else's writing. Put yourself *in loco lectoris*, which means in the place of the reader. Don't be satisfied with your work until it is just as clear, lean, and readable as you can make it.

The following effusions suggest that their authors didn't follow the advice just given. From a geophysical journal: *While wet years are more numerous than dry years, dry years are, on the average, drier than wet years are wetter*. From a trade journal: *The Moenkopi Formation contains estimates of over one billion barrels of oil*. (The formation contains no estimates.) From *Science*: *The earliest insect-pollinated angiosperms seem to have been visited and pollinated predominantly by insects*. These sentences not only reflect on the authors; they also indicate that it isn't wise to depend on editors to clean up your prose. Avoiding absurdity is a do-it-yourself job.

So one of the good reasons for a cooling-off period between writing and revising is that statements that seemed perfectly sensible when you wrote them turn out to lack--shall we say--a full measure of cogency.

Perhaps you have noticed in this book the repeated advice to avoid words that aren't necessary. *Omit needless words*! You could do worse than adopt that little motto as your guide. Why write *in a subaqueous environment* when you just mean *under water*? *The cessation of pronounced negative cratonic activity* simply means *when rapid sinking stopped*. Write *depends on*, not *is dependent upon*; write *before*, not *prior to*. Use short concrete terms, the active voice, and good common sense.

Best of luck!

Chapter 4

ILLUSTRATING

Most papers in earth science require at least a minimum of illustration--for example, a photograph and an interpretative map or diagram. At the other extreme is the paper in which the photographs or drawings are the center of attention and are the main reason the paper was written--for example, an interpretation of subsurface geology by means of seismic cross sections; a study of a region from aerial photographs; or a description of microfossils with several plates of photographic enlargements. Wherever your paper falls in this spectrum of needs, include illustrations only if they supplement or enhance the text. Their purpose is neither to decorate nor to duplicate what you said in words, but to clarify it. Choose your illustrations to help your reader.

If you happen to be a photography enthusiast, or are handy with drafting and lettering equipment, by all means take advantage of these abilities. If not, you may need to call on specialists in your organization, or even on outside professionals. In any case, ask for advice from more experienced colleagues. If you are writing for publication, decide on the journal to which you wish to send your report, and don't hesitate to ask the editor how the illustrations should be handled.

In preparing a thesis or an in-house report, not to be published, you may incorporate photographs and line drawings right in the text. Color prints and hand-colored drawings may be usable under these circumstances. In assembling a report for publication, however, illustrations are kept separate from the text; and, because of expense in reproduction, color is rarely acceptable. An adequate black-and-white print may occasionally be made from an exceptionally good color negative or slide, but don't count on this procedure. Normally it is assumed that illustrations are black-and-white.

PHOTOGRAPHS

A good subject in good light with a well-focused camera should yield a good picture--assuming that the photographer centered on what was wanted and was close enough to show it well. If a camera that takes instant photos is available, it may be used to take a trial picture before the photo that will be used is taken with a conventional camera. The final print may be cropped (trimmed) in order to center the subject or delete undesired material. Pictures should show good contrast, but not areas of brilliant white or solid black. Various processes of retouching are possible--airbrushing, pencil retouching, "dodging" light areas to make them darker, and "burning" dark areas to lighten them. However, you will not recover much detail in this way. Sharp focus is essential.

Be sure the picture has a scale: in "scenics" a person, car, or tree, and in closeups a hammer, coin, or other familiar object. The person or object should be at the side of the view, not at the center. In case you forgot the small object, or it was cropped, include in the caption a statement like "Largest pebble is 2 inches across." Don't give a mathematical scale ("X 1.5"), because the picture may be enlarged or reduced in printing and thus invalidate the ratio. A bar scale is acceptable--drawn on an overlay, not on the photograph itself. Glossy prints in the general range of 5 by 7 inches or 8 by 10 inches are desired: check with the journal editor as to preference. Most pictures will be reduced in size in printing.

Handle all photographs with care! Never use staples or paper clips. Don't write on the back, as pressure marks may show through. Tape to the back of each photo a notation with your name, the title of your paper, and the figure number: also, indicate which direction is up, so the picture won't be printed upside down or sideways. Some editors prefer to have the photos mounted on white cardboard, in which case the identification will go on the back of the mount. If lines, letters, or arrows are to appear on a photograph, to aid the reader, place them on a transparent overlay, which is marked as to precise position by registration points outside the photo itself. Be sure there is adequate contrast between this guide material and the photograph. You can obtain adhesive letters and numerals in a variety of sizes and styles at art stores; or you can do the work with a lettering guide. Cover each photo with a protective sheet of acetate or paper, or put it in a transparent plastic envelope. In mailing, always reinforce with stiff cardboard to prevent bending.

In the margin of your typed text, indicate where you would like to have each figure appear by writing Fig. 1, Fig. 2, etc., and encircling it. The printer will try to see that each figure comes out where you want it.

Specialized techniques are involved in preparing micrographs of thin sections, electron micrographs, enlargements of microfossils, and similar photographic figures. Such reproductions may constitute the very basis of the report in which they appear. Presumably you will not be thrust into these special techniques on your own, but will learn them as needed by working with a person directly involved.

LINE DRAWINGS

Maps, cross sections, and other line drawings should be made larger than their size will be when printed--generally twice as large. Dimensions will be governed by those of the journal page. If possible, prepare drawings to fit upright on the page, rather than sideways. Lines and symbols are sharpened and smoothed by moderate photographic reduction, but if reduction is excessive, letters and patterns may close, or "plug." You can see how your drawing will look when reduced by using a reducing lens, or by having a photographic or photocopy reduction made. Some journals prefer reduced photographic prints rather than the original art work, which is larger and more difficult to handle.

Every drawing should be uncluttered, clear, and readable. As to the map, perhaps the wisest counsel is, don't try to show too much detail. Make one or two preliminary sketches of the whole figure, to arrive at the best arrangement of the map itself; a legend, with symbols and pattern blocks; graphic scale, date, and other supporting information; and a concise title. Locate clearly the lines of any cross sections. And be especially careful to see that all localities, formation names, and other items referred to in the text are on the map and in the right form. Few things infuriate a reader more than to have a reference in the text to "Coldwater Canyon (Figure 1)" and be unable to find this canyon anywhere on the figure; or to find a unit that is referred to in the text as the Snarf Formation appearing on the map as the Snarf Group. Text and map must be in phase!

Cross sections should be set forth in the same terms as on the map and in the text, and should be keyed to the map by index letters at either end. Both horizontal and vertical scales should be given, in graphic form, as by the two bars of an L-shaped figure. The vertical exaggeration must be plainly stated. This is ordinarily done in figures, e.g. V.E. 52X. A more graphic way of stating exaggeration is to place below the cross section another one of the same length with vertical and horizontal scales the same. On such a section you can show only the bottom line and the land surface. This skeleton section will show very nicely how much the section above it has been expanded. It is generally unnecessary to letter a title on a cross section, as this information can readily be given in the printed caption.

TABLES

Include tables only for data that the reader must have in order to evaluate your conclusions. Don't feel that you must list every analysis, age determination, or production figure that your research has generated. If you have much statistical information that is of secondary or incidental interest, place it in an appendix; or include a statement in the text to the effect that you will furnish additional data on request.

Some authors find it advisable, especially in summary papers, to put "Previous Work" in the form of a table. Such a table, to include names, dates, and contributions, can be organized by topic and annotated on the scope and significance of each author's work. This arrangement saves the reader from having to wade through text that consists mostly of citations to the literature.

If you must tabulate, arrange the data in a form as easy as possible to absorb. Information that reads most logically from left to right should be placed that way; arrays of numbers that the reader may want to add up (e.g. percentages) should go in columns. Type the title at the top:

*Table 6.2 Comparison of Chemical Composition of Reef and Interreef Dolomite, Racine Formation, Thornton, Illinois**, and a footnote, if needed, at the bottom:

**The Thornton quarry is situated 5 miles south of Chicago and 5 miles west of the Indiana state line. Analyses are from Lowenstam, Willman, and Swann (46, p. 11).*

Identify units of measurement at the head of the columns, not in a footnote. Be sure readers understand your abbreviations. Avoid repeating *000,000* down a column by subheading the column (*in millions of dollars*) or (*in millions of tons*). Don't place horizontal or vertical rules on your tables; the printer will do that. In your text, refer to tables by number, not by expressions like "in the following table." Printers will do their best to put your tables where you want them, but the table that follows in the typescript may not follow on the printed page.

If you are adept at working with a word processor or an electronic typewriter with carbon ribbon, you can experiment in designing your tables and perhaps do them yourself. Camera-ready copy is less expensive to publish, since it doesn't need to be set in type; it will be free of printer's errors and needs no proofreading. On the other hand, preparation of camera-ready copy requires great care and strict accuracy on your part. Any errors are there for all to see, and there is no question as to whose fault they are.

GRAPHS

Numerical relationships are often more readily apparent on a graph than in a table. Graphs allow comparisons, indicate trends, and make prediction possible (if not always accurate). But their main advantage is simply in being graphic, and thus more quickly assimilated than the equivalent set of numbers. Lines on a standard X-Y diagram (X, horizontal, =abscissa; Y, vertical, = ordinate) can show variations in several sets of data by appropriate curves. Each curve should have a distinctive pattern, and also be identified by label. Area or bar diagrams are especially useful for comparing related categories. Other graphic illustrations are the correlation chart, the ratio graph in which one axis is arithmetic and the other logarithmic, and the pie diagram, favored by treasurers because it so neatly shows where the money comes from or where it goes. Mineral or rock compositions may be shown on a triangular diagram, each apex marking a different component. Boundaries of the fields of stability of the various phases of a system appear on phase diagrams. Triangular and phase diagrams can be considered as variants of the map.

Remember that a graph is a drawing. If you prepare it on cross-section paper, trace the final copy on white paper or tracing cloth. Use a lettering guide for all letters and figures. In many journals, printing will reduce the graph to column width (about 2.5 inches), so keep this in mind when deciding how much information to include and what size of letters to use. The columnar format means that if you have two or more related graphs in a single arrangement, they should be placed one above the other rather than side by side. Captions will be set in print, so they must be typed and kept separate from the drawings. As with maps, be sure that all words on the drawing and in the caption are spelled correctly, and that the terminology used on the graph is the same as that in the text.

CAPTIONS

As the nature of many illustrations is self-evident, their captions don't need to serve as labels. Credit readers with the ability to recognize a photograph; it is not necessary to begin the caption *Photograph of.* If a map or section contains a lettered title, don't repeat it in the caption. Perhaps all the identification the figure needs is Figure *2. Cross section along line A-A' of Figure 1.* Occasionally, discussion in the text may make it unnecessary to put anything more in the caption than the figure number.

But most illustrations require brief explanation. Call attention to features in a photograph that exemplify points made in the text; to aspects of a map that show trends or relationships; to comparisons or contrasts brought out by graphs. Remember that what may be clear or obvious to you may not be so to readers. Be sure that the features mentioned in the caption are in fact on the figure and do in fact show what you say they do.

If you think ahead, you may use the caption for information that would be awkward or tedious to letter on the figure. For example, a section through sedimentary rocks that emphasizes mineral deposits has little room for stratigraphic information, so the caption includes the sentence *All the rock units belong in the upper Precambrian Crystal Springs Formation.* Such brief explanatory statements are appropriate for the caption. They help readers.

Chapter 5

TALKING

Sooner or later the time will arrive when you give an oral paper at a meeting. You are allotted, say, 20 minutes. A projector is available for your slides.

A few members of the audience may be well informed on your subject, but most will be unfamiliar with it. Of those who have read your abstract in the program of the meeting, not many have retained anything you said there. Hence it is wise--indeed, in a 20-minute talk it is essential--to limit yourself to a few major points.

What trends, comparisons, or relationships can you share with this audience? Omit supporting data: neither you nor the audience have time for them. Make it simple. Confused readers can always go back and review; confused listeners cannot. Therefore careful organization and a logical sequence of points are imperative. In the first four or five minutes you should outline the subject, being sure to define any special terms or key phrases. The main part of the talk (say, 10 minutes, with lights out if you have slides) will contain the three or four main points that you wish to make, with your comments and interpretation. The final few minutes, with lights on, will give you time to summarize and to repeat the conclusions that you want the audience to remember.

How to prepare for such a presentation? One way, obviously, is to write out your talk and read it. This is all right if you are a good reader, and if you know your talk well enough to give your audience more eye contact than a few fleeting glances between paragraphs. If you read, however, you have to stick to the script; when a slide comes on the screen, you may lose your place on the precious pages if you turn to comment on the slide informally. Another possibility is to learn your talk and recite it from memory, but this tends to emphasize recitation at the expense of content. Much better is to have firmly in mind the major points that you wish to make, and the slides that go with them, and then give the talk informally with the aid of notes or a topic outline to ensure that you stay on the track.

However you prepare for your talk, there is one absolute requirement: *rehearse*. The few minutes allowed mean that timing is essential; slides must be phased in; your main points must be made. Seclude yourself in a room, have your notes in hand, stand up, and give your talk--again and again, until you know it will fit your time slot on the program and will give the audience your message. Knowing that you will finish in the allotted time will relieve you of one of a speaker's main worries: that the 2-minute warning light will flash on when he or she is only part way through. Staying on time will also endear you to the session chairman and the audience. Repeated rehearsal beforehand is the only way to guarantee this desirable result.

Another morale-builder--and a good way to leave your audience impressed and maybe even with some idea of what you said--is a good set of slides. The rule that applies to your whole talk--keep it simple--applies to each slide. One slide, one idea. Don't ask your audience to absorb a detailed geologic column or a set of tabulated figures in the few seconds when the slide is on the screen. *Illustrations prepared for print are useless*: too much information, figures and letters too small, lines too thin. Every slide must be specially prepared. Color is good but not essential. The main object is to present in graphic style a single point, relationship, or conclusion per slide. Titles are unnecessary: what the slide shows is generally clear or you can explain it. Use letters big enough to be read from the back of the room; omit legends, putting the necessary labels right on the figures; simplify and generalize geologic maps. And never show typed columns of numerical data! These belong in the printed version of your paper, not on the screen in an auditorium.

By far the best way to gain first-hand information on the techniques of giving an oral paper is to attend a meeting where such papers are presented and to sit through several of them. You will almost certainly encounter examples of how not to do it--a speaker mumbling about statistical data while the screen is full of columns of unreadable figures, another speaker ignoring the time warning and encroaching on the time of the following speaker, and so on. On the other hand you will see and hear several talks that reflect well on speakers and on their regard for the audience--a clear, understandable message backed by good slides, concluding neatly at the end of the allotted time slot. Over the years, several cynical geologists have suggested that, at each meeting, special recognition be given to the speaker who has presented the worst set of slides, the recipient being identified by audience reaction. If this suggestion is ever implemented, and you are on the program, I hope that your paper will never even be considered for such a dubious honor.

Chapter 6

FURTHER INFORMATION

American Geological Institute 1986. *Writer's Guide to Periodicals in Earth Science*, 2nd edn. Alexandria, Virginia: American Geological Institute. Information on more than 50 geological journals for the first-time writer--subjects considered, format required, and procedures for submitting papers.

Bishop, E.E., E.B. Eckel, and Others 1978. *Suggestions to Authors of the Reports of the United States Geological Survey*, 6th edn. Washington: U.S. Government Printing Office. A comprehensive guide to the Survey's practice, covering report writing, preparation of illustrations, stratigraphic nomenclature, and a variety of other topics.

Clifton, H.E. 1978. "How to Keep an Audience Attentive, Alert, and Around for the Conclusions at a Scientific Meeting." *J. Sed. Petrol.* 48, 1-5. Reprinted in *GSA News & Information* 7, 88-90, 1985. The title of this paper is accurate. Required reading for anyone planning to give a talk.

Cochran, W., P. Fenner, and M. Hill (eds.) 1984. *Geowriting: a Guide to Writing, Editing, and Printing in Earth Science*, 4th edn Alexandria, Virginia: American Geological Institute. This is just what its title says. Up to date, broadly inclusive, and notably unstuffy.

Day, R.A. 1979. *How to Write and Publish a Scientific Paper*. Philadelphia: ISI Press. Witty counsel on all aspects of the subject. Emphasis is on biology but earth scientists will find much of value.

Heron, D.E. (ed.) 1986. *Figuratively Speaking: techniques for preparing and presenting a slide talk*. Tulsa, Oklahoma: American Association of Petroleum Geologists. Written for anyone who speaks in public, especially about geology.

Hill, M. and W. Cochran 1977. *Into Print: a Practical Guide to Writing, Illustrating, and Publishing*. Los Altos, California: William Kaufmann, Inc. A readable handbook on all aspects of the subject.

Hunt, C.B. 1964. "Some Suggestions on Writing a Thesis." *J. Geol. Education 12, 130-1*. The abstract starts off "A well-prepared thesis requires a certain lack of clarity" and goes on from there. The irrepressible Dr. Hunt is in good form.

Landes, K. K. 1966. "A Scrutiny of the Abstract, II." Bull. Am. Assoc. Petrol. Geol. 50, 1992. On this one page, Landes tells you what you need to know in order to prepare the informative abstract that the journals require.

Murray, M.J., and H. Hay-Roe 1986. *Engineered Writing*, 2nd edn. Tulsa, Okla.: PennWell Books. A "user-friendly manual" especially for writers in the corporate field. Much good advice on preparing reports for management.

Pratt D., and R. Ropes 1978. *35-mm Slides: A Manual for Technical Presentations*. Tulsa, Oklahoma: American Association of Petroleum Geologists. An indispensable guide to the preparation of slides. Copiously illustrated, well organized, highly practical.

Rayner, D.H. 1982. English Language and Usage in Geology: A Personal Compilation. Transaction Leeds Geological Association, Special Issue, 30 p. Good help for "geologists who are having to write up their observations or research," especially "the less experienced who are composing from professional necessity rather than from inclination."

Shinn, E.A. 1981. "Make the Last Slide First." *J. Sed. Petrol.* 51, 1-6. Examples of good and bad slides, and advice on how to make good ones.

Vanserg, N. 1952. "How to Write Geologese." *Economic Geology*. 47, 220-3. Nicholas Vanserg is the pseudonym of the late Hugh McKinstry, longtime professor of geology at Harvard University. He gives some examples of how not to write English, with suitably acerbic comments.

Chapter 7

THREE SAMPLES

Here are three samples of good writing in earth science. In no way remarkable, they simply illustrate the kind of clear, informative, readable prose that readers have a right to expect. After considerable experience, and with good editorial help, you should be able to write as well or better.

From *Quaternary Geology of the San Andreas Fault Zone at Point Reyes National Seashore, Marin County, California,* by N. Timothy Hall and David A. Hughes, California Division of Mines and Geology Special Rept. 140, 1980:

> The San Andreas fault zone in Marin County is expressed topographically as a remarkably straight trough, from 0.8 to 1.6 kilometers wide and 21 kilometers long, that separates the Point Reyes Peninsula on the west from "mainland California." The drowned ends of this trough form Bolinas Lagoon on its southeastern end and Tomales Bay on the northwest. The predominant topographic features of the subaerial portions of the rift valley are low ridges separated by valleys or depressions, all of which are generally elongated parallel to the fault zone. These ridges do not exceed 5 kilometers in length and 50 meters in height. Some end by wedging out; others are replaced along the same trend by linear valleys. Closed depressions containing swamps or ponds are common features of both the ridges and the intervening valleys. G.K. Gilbert counted 47 such sag ponds between Bolinas Lagoon and Lagunitas (Papermill) Creek during his study of the fault zone after the 1906 earthquake (Lawson, 1908, p. 31). Because he found that the surface changes caused by the 1906 faulting nearly always tended to increase the heights of the ridges and the depths of the val-

> leys, he concluded that these linear features were of fault origin. The complex and often anomalous drainage patterns within the rift zone also support a tectonic origin for most of the ridges and valleys. The two major subsequent streams draining the rift valley are Olema Creek, which flows northwestward into Tomales Bay, and Pine Gulch Creek, which flows southeastward into Bolinas Lagoon. For a distance of 3 kilometers these axial streams flow parallel to each other in opposite directions and are separated only by low ridges. In general, the ridges within the rift zone are independent of the drainage. In part, they deflect or localize the drainage; in part, the streams cut across the ridges and create water gaps. As might be expected, streams often follow the linear valleys between the ridges, but they also commonly cut across these valleys.

From *Chemical Weathering of Basalt and Andesites: Evidence from Weathering Rinds*, by Steven M. Colman, U.S. Geological Survey Professional Paper 1246, 1982:

> Compositional variation in minerals appears to affect the location and the rate of weathering. Many of the rocks examined contain plagioclase that has normal, reversed, or oscillatory zonation; the more calcic zones in these minerals altered first and more rapidly. In pyroxenes, varying degrees of alteration due to compositional variation were only rarely observed. Compositional zonation in olivine, if present, did not affect the location or the rate of weathering.
>
> As time passes or as weathering becomes more severe, minerals become encased in an increasingly thick sheath of weathering products. Hence, in time this sheath may impede the movement of weathering solutions to and from the remnants of primary minerals. In extremely weathered portions of some weathering rinds, nearly fresh remnants of primary minerals appear to be protected in this manner by sheaths of weathering products.

From *Geology of the Nonmetallics*, by Peter W. Harben and Robert L. Bates, New York: Metal Bulletin, Inc., 1984:

> Diatomite deposits commonly contain more or less clay, volcanic ash, silt, and other impurities, and there are all gradations from diatomite through diatomaceous shale to clay shale and siltstone. Diagenetic changes or metamorphism may destroy the open porous structure of the diatomite and produce porcelanite or opaline chert. Commercial diatomite generally contains 86 to 94% silica, the remainder being chiefly alumina and alkalies from included clay. As mined, the rock contains 50% or more of moisture. Beds range

in thickness from a few inches to several hundred feet; the bedding may be finely laminated or massive.

INDEX